Oleynik Pavel Petrovich

Basic standards of corporate information systems:

Oleynik Pavel Petrovich

Basic standards of corporate information systems:

MPS, MRP, MRP II, ERP, CSRP, ERP II

ScienciaScripts

Imprint
Any brand names and product names mentioned in this book are subject to trademark, brand or patent protection and are trademarks or registered trademarks of their respective holders. The use of brand names, product names, common names, trade names, product descriptions etc. even without a particular marking in this work is in no way to be construed to mean that such names may be regarded as unrestricted in respect of trademark and brand protection legislation and could thus be used by anyone.

Cover image: www.ingimage.com

This book is a translation from the original published under ISBN 978-3-8433-1440-4.

Publisher:
Sciencia Scripts
is a trademark of
International Book Market Service Ltd., member of OmniScriptum Publishing Group
17 Meldrum Street, Beau Bassin 71504, Mauritius
Printed at: see last page
ISBN: 978-620-3-18023-7

Contents

Introduction

Implementation of a unified corporate information system (CIS) is an indispensable condition for the functioning of a modern enterprise in any field of activity. Planning and implementation of a large information system, in its essence, acts as a reform of the enterprise management system. Therefore, correctly prepared plan will avoid many problems that arise during the implementation of the system and its subsequent operation. Change in the management system, first of all, is connected with application of the newest methods of work with information and data. Reforming relates to business process management, planning, budgeting, controlling, etc. The availability of a large number of ready-made solutions contributes to the difficulty of choosing the optimal system.

Knowledge of the basic standards governing the functional features of the EIS contributes to minimizing costs when implementing a software product and allows you to best meet the needs of a particular company. This book is devoted to the study of the fundamental features and basic standards of CIS.

General information about corporate information systems

Nowadays, no large enterprise can exist and develop without a highly effective management system based on the latest information technologies. Constantly changing market requirements, huge flows of information of scientific, technical, technological and marketing nature require the personnel responsible for the strategy and tactics of high-tech enterprise development to make quick and accurate decisions aimed at obtaining maximum profit at minimum cost. This is partly why the idea of building corporate information systems (CIS) has recently become so popular [1-2].

Corporate Information System (CIS) is a defined set of methods and solutions used to create a single information space for management and operation of the company.

Despite the fact that the concept of corporatism implies the presence of a fairly large, as a rule, geographically distributed information system, it is still quite legitimate to consider any information systems here, regardless of the architecture implemented in them, if these IS are designed to automate the main business processes of the enterprise. Creating a unified information system that allows to automate all key business processes and effectively manage (take into account) the limited resources (money, time, production capacity, etc.) in the enterprise is the main goal of the implementation of CIS.

1.1 Requirements for the EIS

Historically, a number of the following requirements for corporate information systems have been formed [1-3]:

1. Systematicity;
2. Complexity;
3. Modularity;

4. Openness;

5. Adaptability;

6. Reliability;

7. Safety;

8. Scalability;

9. Mobility;

10. Easy to learn;

11. Support at the stage of implementation and support from the developer.

Let us consider each of these requirements in more detail. First of all, the IMS must meet the requirements *of comprehensiveness* and *consistency*. It covers all levels of management from the corporation as a whole, taking into account branches, subsidiaries, service centers and representative offices to the shop, area and a specific workplace. The entire production process, from the point of view of informatics, is a continuous process of generating, processing, changing, storing and distributing information. Each workplace - whether it is the workplace of an assembler on an assembly line, an accountant, a manager, a storekeeper, a marketing specialist or a technologist - is a node that consumes and generates certain information. All such nodes are connected by information flows, embodied in the form of documents, messages, orders, actions, etc.

The next requirement imposed on the EIS is *the modularity* of construction. This requirement is also very important in terms of system implementation, as it allows to parallelize, facilitate and, accordingly, accelerate the process of installation (installation of the application on users' workstations), personnel training and launch of the system in commercial operation. Moreover, if the system is not created for a particular production, but is purchased on the market of ready-made software solutions, the modularity allows to exclude from the delivery the components that do not fit into the infological model of a particular enterprise, or without which at the initial stage can be done, which reduces the cost of implementing the EIS.

Since no typical information system implemented at a real enterprise can be exhaustively complete, and due to the fact that the functioning enterprise may have components of other CIS already in operation and proven to be useful, the next defining requirement is *openness.* This requirement becomes particularly important if we consider that automation is not limited to management, but also covers such tasks as design and maintenance, technological processes, internal and external document management, communication with external information systems, security systems, etc.

Any enterprise does not exist in a closed space, but in a world of constantly changing supply and demand, requiring a flexible response to the market situation, which can sometimes be associated with a significant change in the structure of the enterprise and the range of products or services provided. In addition, in an economy in transition, the law is fluid and dynamic. Large corporations, in addition, may have territorial divisions located in the jurisdiction of other countries or free economic zones. This means that the CIS must be *adaptive*, i.e. flexible to different legislation, have interfaces in different languages, be able to work with several different currencies simultaneously. Without adaptivity the system is doomed to a short existence, during which the cost of its implementation will not be recouped. It is desirable that in addition to customization tools the system had also the means of development (functionality expansion) - tools with which programmers and the most skilled users of the enterprise could independently create the necessary components (forms, reports, etc.), organically embedded in the system.

When the EIS is operated in industrial mode, it becomes an indispensable component of a functioning enterprise, capable of stopping the entire production process and causing huge losses in case of an emergency failure. Therefore, one of the most important requirements for such a system is the *reliability* of its operation, which implies the continuity of the system as a whole, even in case of partial failure of its individual elements due to unforeseen and insurmountable causes (operational errors).

Security is extremely important for any large-scale system containing large amounts of information. The security requirement includes several aspects [1]:

1. *Protection of data from loss*. This requirement is implemented mainly at the organizational, hardware and system levels.

An application system, such as an automated control system (ACS), does not necessarily need to contain data backup and recovery facilities. These issues are handled at the operating environment (or DBMS) level.

2. *Maintaining data integrity and consistency*. The application system must track changes to interdependent documents and provide version and generation management of data sets.

3. *Preventing unauthorized access to data within the system*. These tasks are solved comprehensively by both organizational measures and at the level of operating and application systems. In particular, applied components should have developed administrative tools that allow to limit access to data and system functionalities depending on user status, as well as to monitor user actions in the system.

4. *Preventing unauthorized access to data from the outside*. Solving this part of the problem falls mainly on the hardware and operating environment of the CIS and requires a number of administrative and organizational measures.

Successfully operating enterprise and receiving sufficient profit tends to grow, the formation of subsidiaries and branches, which in the operation of CIS may require an increase in the number of automated workstations and the volume of stored (processed) information. In addition, for companies such as holdings and large corporations should be able to use the same data management technology as the parent enterprise, and at the level of any, even a small member firm. This approach to information management puts forward the requirement *of scalability*.

At a certain stage of an enterprise's development, increasing demands on system performance and resources may determine the transition to a newer hardware and software platform. To ensure that such a transition does not entail a radical change in the management process and unjustified investments in the purchase of more powerful application components, it is necessary to meet the *mobility* requirement.

Ease of learning is a requirement that includes not only the presence of an intuitive graphical interface of the programs, but also the availability of detailed and well-structured documentation, opportunities for staff training on specialized courses and internships for responsible professionals in companies related to the profile, where the system has been successfully implemented and operated.

The next requirement **is *support from the developer***. This concept includes a number of opportunities, such as receiving new versions of the software for free or at a significant discount, getting additional methodological literature, round-the-clock hotline consultation, receiving information about other software products of the developer, the opportunity to participate in seminars, scientific conferences of users and other events held by the developer or user groups, etc. Only a large company that has been on the software market for a long time and has a rather clear perspective for the future is able to provide such support to the user.

Another requirement is ***support***. In the process of operation of complex software and hardware complexes there may be situations that require prompt intervention of qualified personnel of the company-developer or its representative on site. Maintenance includes a specialist on-site visit to the customer to eliminate the consequences of accidents, methodical and practical assistance if necessary to make changes to the system that are not of a radical restructuring or new development. It also includes installation of new versions of the software, received from the developer free of charge by employees of the authorized accompanying organization or by the developer himself.

1.2 Standards governing the functionality of the EIS

The modern market requires that all products meet generally recognized quality standards, which concern not only the quality of the final product displayed in the market, but also the entire process of manufacturing this product, from the selection of suppliers of components to the service of the finished product. At present the set of standards for enterprise quality system developed by ISO (International Standards Organization) is spread all over the world. This set of standards has a common name ISO 9000 (ISO 9000). Part of the standards of this complex regulates the functional elements present in the QIS.

The introduction and maintenance of the enterprise quality system in accordance with the standards of the ISO 9000 family involves the use of software products of at least three classes [1-

3]:

1. Integrated enterprise management systems (automated information systems to support management decision-making), AISPPR;

2. Electronic document management systems;

3. Products that allow you to create functional models of the organization, to analyze and optimize its activities (including the lower level class systems and CAD, data mining products, as well as software, focused exclusively on the preparation and maintenance of quality systems in accordance with ISO 9000).

This does not mean that every enterprise claiming to comply with the ISO

9000 quality system must necessarily have a unified corporate information system. But managing the huge amounts of data that circulate in an enterprise without a corporate information system would be very difficult. The presence of a corporate information system allows you to maintain the required ISO 9000 level of quality with less cost for documentation and decision-making. Thus, the implementation of ISO 9000 quality system and the introduction of a corporate information system at the enterprise are interrelated. This allows the following (functional) definition of the corporate information system [1]:

Corporate Information System (CIS) - a set of information systems of individual departments of the company, united by a common document flow, such that each system performs part of the tasks of decision-making management, and all systems together ensure the functioning of the enterprise in accordance with ISO 9000 quality standards.

Throughout the history of corporate information systems, many standards governing them have been developed. This or that system, which by its functionality meets a certain standard, belongs to a class of this standard and is automatically called the same as it.

Figure 1 shows the evolutionary path of the most well-known standards.

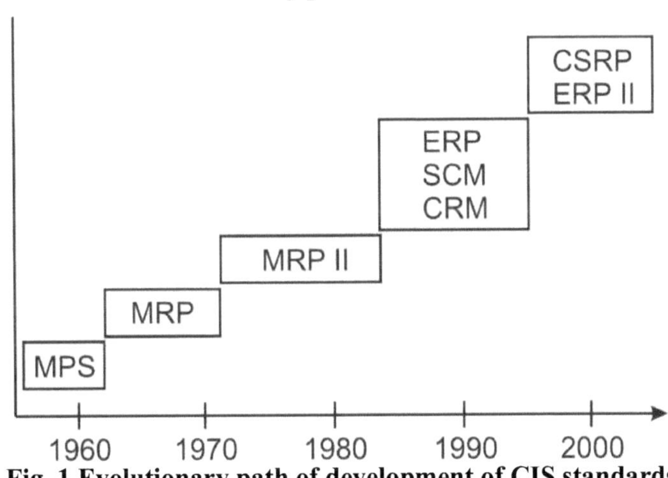

Fig. 1 Evolutionary path of development of CIS standards

The original standard, which appeared in the late 50's - early 60's, was the MPS (Master Planning Scheduling) standard, designed to produce the main production plan. Based on the data on the state of demand, plans for the production of final products were developed [1-2].

For the purpose of optimal production management in the mid 60's the principles of enterprise material inventory management were formulated. These principles formed the basis of MRP (Material Requirement Planning) class systems.

needs). These systems are designed to calculate the possibility of fulfilling a new order by a specified deadline at a given production load. Under conditions of impossibility to execute the given order by the given deadline, the system is able to answer the question what it will cost to execute the new order, if the customer still insists on the original deadline.

Then came the MRP II class systems (Manufacturing Resource Planning), the main essence of which is reduced to forecasting, planning and production control, which are carried out throughout the cycle, from the purchase of raw materials to the shipment of the final product to the consumer. In general, they provide the solution of enterprise activity planning tasks in physical units and financial planning in monetary terms [4].

The next stage in the development of CIS was the emergence of ERP (Enterprise Resource Planning) systems from the late 80s. These systems cover all financial, economic and production activities of the enterprise [5-6]. They have such requirements as: centralization of data in a single database, near-real-time mode of operation, preservation of a common management model for enterprises in any industry, support of geographically distributed structures, work in a wide range of hardware and software platforms and DBMS. Other important requirements for ERP-systems are: the possibility of using graphics, CASE-technologies for further development of the system, support of client-server architecture and their implementation as open systems. With the proper implementation and operation of such systems, the efficiency of business processes of the enterprise increases, which gives a competitive advantage for further development. However, by improving the internal structure, the enterprise does not increase the efficiency of interaction with counterparties (external organizations and firms).

The next stage in the development of CIS is focused on integrating the activities of customers and partners of the enterprise into its internal system and is called ERP II (Enterprise Resource and Relationship Processing). Internet provides an opportunity for an enterprise to interact with all its counterparties in a completely new environment, allowing contact directly with the consumer by type B2C (Business-to-Consumer) or/and with business partners by type B2B (Business-to-Business) [6, 10].

In order for ERP II system to be applicable to e-commerce and business it is necessary to create CRM (Customer Relationship Management) applications, as well as additional software (software) of the intermediate layer [7-8]. Such software is called EAI (Enterprise Application Integration). EAI systems provide the following functions:

1. E-commerce;
2. Supply chain management;
3. Application Access Services;
4. Virtual Marketplaces.

An ERP II system with CRM and EAI products is called an XRP system, i.e. an Extended ERP system. With its help, it is possible to share data circulating among various corporate applications in real time. In terms of classification this system approaches the next generation of CIS - CSRP standard systems (Customer Synchronized Resource Planning - resource planning with customer). Systems of this class allow integration of processes both within a corporation and outside it into a single whole [9].

2. MPS standard

In the late 50's and early 60's, due to the growing popularity of computing systems, the idea arose to use their capabilities to plan the activities of the enterprise, including the planning of production processes. The need for planning was due to the fact that most of the delays in the production process were associated with delays in the supply of individual components. As a result, in parallel with a decrease in production efficiency, an excess of materials arriving on time or ahead of schedule appears in the warehouses. In addition, due to the imbalance in the supply of components, there are additional complications in recording and tracking their status during production. That is, it is impossible to determine which batch a given component belongs to in an already assembled finished product (product).

2.1. Scheme of the MPS-system functioning

The first business management standard was MPS (Master Planning Scheduling), or the volumetric scheduling standard. The idea of the standard is obvious and is presented in Fig. 2. First the sales plan was formed, i.e. the sales volume was determined with the distribution by calendar periods. Based on the sales plan, a replenishment plan is formed by production or purchase, and the financial results are evaluated for different periods, which are used as planning periods or financial periods [1-2].

This standard optimally describes the model of functioning of a relatively small trading enterprise with a fairly simple production scheme.

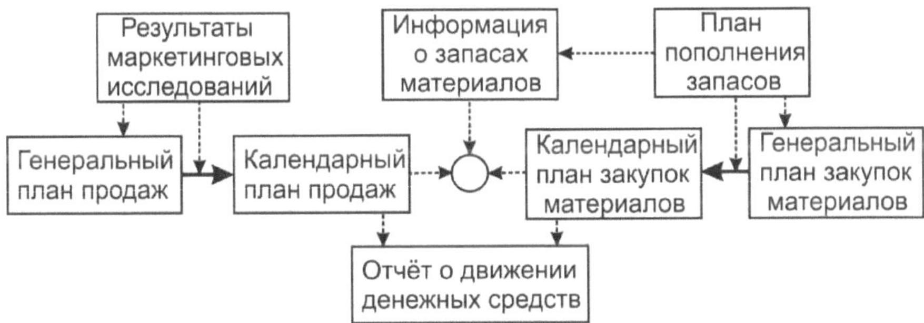

Fig. 2 MPS Scheme of the EIS functioning

If an enterprise is developing dynamically, a number of difficulties arise. The first problems arise with logistics management, because the formed order for the necessary materials may not arrive at the time that was previously planned.

One of the most difficult problems arising in the formation of an order was the problem of forecasting the necessary volume and delivery time, which largely depends on the supplier's capabilities. Consequently, it is necessary to forecast demand for a long time ahead, take into account the duration (and often the season) of production and the need for storage space. In this case, the volume of material orders also can not be expressed in arbitrary numbers. Such problems arise with wholesale (large) sales.

2.2. *Static inventory management*

Small wholesale and retail sales also have their own peculiarities. For example, it is often unacceptable that the   sale of goods

daily demand, because it will lead to the departure of the customer to a

competitor

(a neighboring store, wholesale warehouse), where the product is always

in stock. As a result there is a Safety Stock

in the amount of, for example, a daily requirement. The concept of Safety Stock

is widely

used in manufacturing to ensure

21

continuous production process.

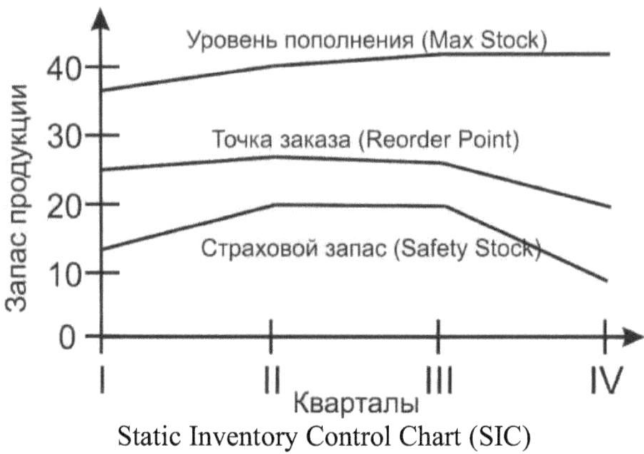

Static Inventory Control Chart (SIC)

Fig. 3 Main parameters of static inventory management

Further study of the dynamics of stocks using statistical methods SIC (Statistical Inventory Control - statistical inventory control) leads to two more concepts - the Order Point (Reorder Point), which determines the level of inventory, when the planned stock below which it is necessary to make or plan an order to the supplier and the replenishment level (Max Stock) of inventory in stock, which is the amount of goods above which it is not recommended to increase the level of inventory. Fig. 3 shows the described parameters affecting the dynamics of stocks [1-2].

Note that the values of the described parameters are dynamic, as the replenishment order must be formed in advance, taking into account the time of delivery, and the volume of delivery may not fit into the planned replenishment level. Dynamism also occurs when taking into account seasonal changes in the main SIC parameters - obviously, the insurance stock of an extensive assortment of soft drinks is quite significant in summer, and in winter the absence of a full assortment will remain unnoticeable to the end consumer. Determination and fixation of such fluctuations is a subject of serious statistical research. Modern information management systems have built-in statistical analyzers of at least the simplest type or autonomous external subsystems that allow to perform such an analysis.

2.3. Ways of presenting the product specification

Even more serious problems began to arise with the increasing complexity of production and the production of complex products, the number of parts in which is measured in thousands, and the assembly of which was carried out on several conveyors.

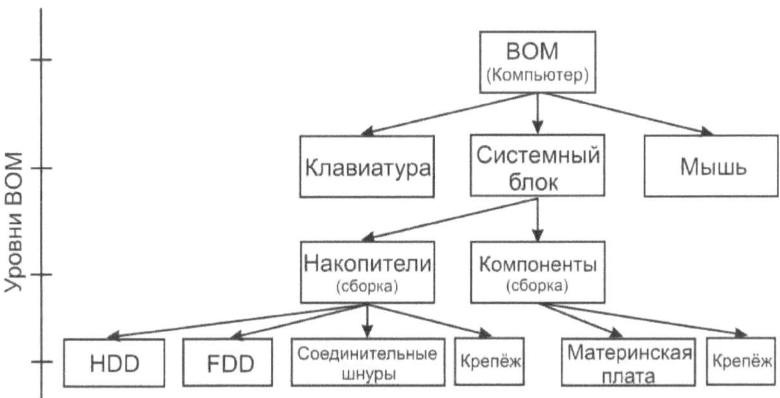

Fig. 4 Hierarchical representation of the product specification levels

This gave rise to the concept of an assembly, subassembly, unit - that is, a component, part, or some constituent part of the final product prepared on an auxiliary assembly conveyor before being installed into the finished product on the main conveyor. Typical examples of such products are engine, chassis and bodywork in mechanical engineering. The products produced during this kind of assembly operations can be represented in the form of tree-like structures (Fig. 4),

The BOM (Bill Of Material) is a generalized term [1-2].

Different levels of PTO can contain the same product items, such as "Fasteners" at different levels of the assembly specification of a computer. The unbundling from the above tree structure results in the linear structure required for the formation of an order for the purchase of materials and presented in Fig.

5.

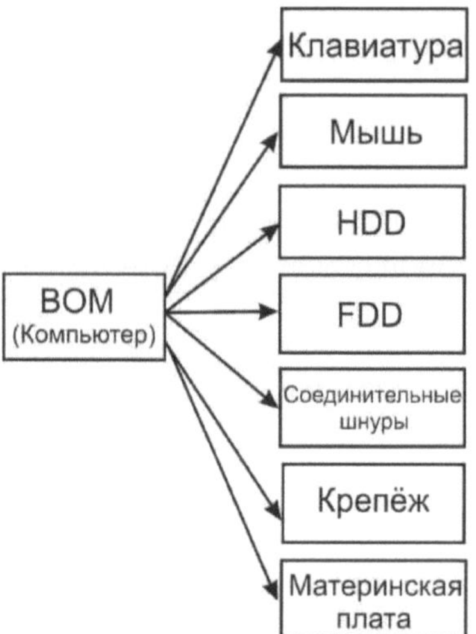

Fig. 5 Linear representation of the product specification

Note that a fastener occurs only once in the line list, because it represents one unique product, just like in the order.

As a result, the inventory management problems described above have become much more complex. In addition to basic components, there were problems with assemblies, which could be produced in a single assembly process, at auxiliary production facilities with intermediate storage of unfinished products (or/and assemblies), or on a subcontracting basis "on the side," and the same

an assembly or subassembly, such as an engine, can either be ordered or manufactured. The requirements for the accuracy of the timing of delivery of this kind of component have become an order of magnitude higher than before for basic components. Thus, there was an urgent need to develop a new standard.

3. MRP standard

As a result of the partial solution of these problems, a methodology of production planning (mainly assembly or discrete), which proposes to form an order for components and assemblies, based on the needs of the volume-calendar production plan, called MRP (Material Requirement Planning).

The implementation of a CIS, working according to the MRP methodology, is a computer program that allows to optimally regulate the supply of components for the production process, controlling the stock in the warehouse and the production technology itself. The main task of MRP is to ensure the availability of all required materials and components at any time within the planning period, along with the possible reduction of permanent stocks and, therefore, the unloading of the warehouse. Before describing the structure of MRP itself, it is necessary to introduce a short list of its basic concepts [1-2]:

- *Materials* are all the raw materials and individual components that make up the final product.

- *MRP program* is a computer program that works according to the algorithm governed by the MRP-methodology. Like any program, it processes data files (input elements) and forms result files on their basis.

- *The material status* is the main pointer to the current material status. Every single material at a particular moment in time has a status within the MRP system, which determines whether it is in stock, reserved for other purposes, present in current orders, or whether an order for it is only planned. Thus, the status of the material unambiguously describes its degree of readiness to be put into the production process for the manufacture of the final product.

- *Material safety stock* is necessary to maintain the production process in case of unforeseen and unrecoverable delays in material supply. Ideally, when the supply mechanism is flawless, MRP methodology does not require a safety stock or its amount is set differently for each specific case, depending on the current situation with the receipt of materials.

• ***The material need*** in the MRP program is a certain quantitative unit that reflects the need to order this material at a certain point in time (during the planning period). A distinction is made between the concepts of full material need, which determines the necessary quantity required to send to production, and net material need, the calculation of which takes into account the availability of all safety and reserve stocks of this material. The order in the system is automatically formed when there is a net need other than zero.

The planning process includes functions for the automatic creation of order projects for the purchase and/or in-house production of the necessary materials. In other words, the MRP system optimizes the delivery time of components, thereby reducing the cost of production and increasing its efficiency. The main advantages of using the MRP system in production are [1-2]:

• Guarantee of availability of the required components and reduction of time delays in their delivery, and, consequently, increase in the output of finished products without increasing the number of jobs and loads on the production equipment.

• Reducing manufacturing defects in the assembly process of finished products arising from the use of unacceptable components.

• Streamlining of production by controlling the status of each material, allowing the unambiguous tracking of its entire conveyor path, from order creation to its position in the already assembled finished product. Thanks to this, complete reliability and efficiency of production accounting is achieved.

All these advantages stem from the very concept of the MRP standard, which is based on the principle that all materials and components, parts and blocks of a finished product must arrive in production at the same time, at the planned time, to ensure the creation of the final product without additional delays. The MRP system speeds up the delivery of those materials that are needed in the first place at that moment. It is necessary to avoid the situation when a delivery of one of the materials is delayed and production has to stop even if all the other components of the final product are available. The main purpose of the MRP system is to form, control and, if necessary, change the dates of receipt of the required materials so that all of them arrive at the same time.

3.1. Input parameters and results of the MRP system

In practice, the MRP system is a CIS, the inputs whose parameters and results are shown in Fig. 6 [1-2].

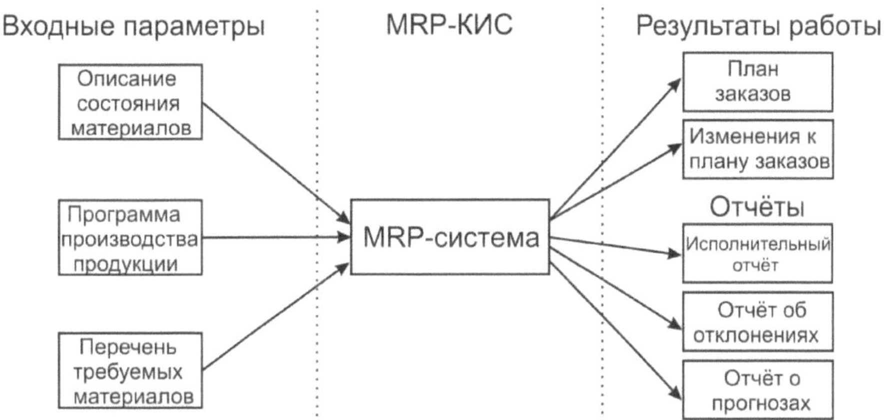

Fig. 6 Structure diagram of MRP-system functioning

The main input elements of the MRP system are:

- The Inventory Status File is the main input element of the MRP program. It should contain the most complete information about all materials and components necessary for the production of the final product. This element indicates the status of each material, determining whether it is available in the production shop, in the warehouse, in the current orders or its order is only planned, as well as a description of its stocks, location, price, possible delays in delivery, details of the suppliers. Information on all of the above items should be specified in the context of each material involved in the production process.

- ***The*** Master Production Schedule is an optimized time schedule for the production of the required batch of finished goods in a planned period or range of periods. First a trial production program is created, which is then tested for feasibility by additionally running it through the CRP (Capacity Requirements Planning) system, which determines whether there is enough capacity to implement it. If the production program is found to be feasible, it is automatically transformed into the main program and becomes an input element of the MRP system.
- The Bills Of Material File is a list of materials and their quantities required to produce the final product. It also contains a description of the structure of the final product, i.e. it contains complete information on the technology of its assembly. It is extremely important to maintain the accuracy of all the entries in this element and to correct them accordingly whenever changes are made to the structure and/or the production technology of the final product.

As a rule, each of the above-mentioned input parameters is a data file used by the MRP program. Currently, MRP systems are implemented on a wide variety of hardware platforms and are included as modules in most financial and economic systems. The main results of an MRP system are:

- The Planned Order Schedule defines how much of each material is to be ordered at each considered time period during the planning period. The order plan serves as a guide for further work with suppliers and, in particular, determines the production program for internal component production, if any.
- Changes In Planned Orders are modifications to previously planned orders. A number of previously placed orders may be cancelled, modified or delayed, or rescheduled.

The following auxiliary reports are also generated:

• The Performance Report is the main indicator of the correctness of the MRP system operation and alerts the user about critical situations in the planning process, such as the complete exhaustion of the safety stocks for individual components, as well as about all arising system errors in the process of the MRP program operation.

• *The* Exception Report is designed to inform the user in advance of time slots during the planning period that require special attention and where external management intervention may be necessary. Typical examples of situations reflected in this report could be unexpectedly late orders for parts, excess parts in stock, etc.

• *A* Planning Report is information used to make projections about possible future changes in the volume and characteristics of products, derived from an analysis of the current production process and sales reports. The Forecast Report is also used for long-range planning of material requirements.

3.2. Algorithm of MRP system operation

The algorithm for any MRP system includes the following key steps:

1. The MRP system analyzes the production program adopted at the input and determines the optimal production schedule for the planned period.

2. All materials that are not in the production program, but are present in current orders, are included in the planning as a separate item.

3. Based on the approved production program and orders for components not included in it, for each individual material the total requirement is calculated in accordance with the list of components of the final product.

4. Based on the total (calculated) demand, taking into account the current material status, a net demand is calculated for each time period and for each material, using the formula below. If the net material requirement is greater than zero, the system automatically creates a material order.

Net Need = Total Need - Quantity Available - Insurance Stock - Reserved

5. All orders created earlier than the current planning period are reviewed and, if necessary, changes are made to the order plan to prevent premature deliveries and delays from suppliers.

As a result of the MRP-program operation a number of changes are made in existing orders and, if necessary, new ones are created to ensure optimal dynamics of the production process.

These changes automatically modify the material status description, as the creation, cancellation or modification of an order respectively affects the status of the material to which it belongs. The program results in a plan of orders for each individual material for the entire planning period, the enforcement of which is necessary to support the production program.

The use of the MRP system for planning production requirements allows to optimize the arrival time of each material, thereby significantly reducing storage costs and facilitating production accounting. However, there is a difference of opinion regarding the use of safety stock for each material. Proponents of the use of an insurance stockpile argue that it is necessary because the delivery mechanism is not sufficiently reliable. Opponents of the use argue that the absence of an insurance stock is one of the central features of the MRP concept, because the MRP system must be flexible with respect to external factors, making timely changes to the order plan, in case of unforeseen and unrecoverable delays in delivery. In real life, the second point of view is true only when planning the needs of production of items for which demand is predictable and controllable. For Russian conditions, when delays in deliveries are the norm, it is advisable to apply planning taking into account the safety stock, the volumes of which are determined in each case separately.

Despite having a number of advantages, MRP systems have the following disadvantages:

- Significant amount of calculation and preprocessing of data;

- Rapidly increasing logistics costs for order processing and transportation as the firm seeks to reduce inventory or switch to small orders with high order frequency;
- Insensitivity to short-term changes in demand.

4. MRP II standard

MRP systems were formed on the basis of the approved production program a plan of orders for a certain period, which is clearly not enough for many enterprises. In order to increase the efficiency of planning in the late 70's the idea of implementing a Closed Loop in MRP-systems was proposed. The idea was to consider a wider range of factors in planning, by introducing additional functions. To the basic functions of production capacity planning (CRP, Capacity Requirements Planning) and material requirements planning (MRP) it was proposed to add a number of additional functions, such as control of conformity of the quantity of manufactured products to the quantity of components used in the assembly process, making regular reports on order delays, on product sales volumes and dynamics, on suppliers, etc. [1-2, 4].

The term "closed loop" reflects the main feature of the modified system, which is that the reports generated during its operation are analyzed and taken into account in further planning stages, changing, if necessary, the production program and the order plan. In other words, the additional functions provide feedback in the system, providing planning flexibility in relation to external factors such as demand levels, the status of suppliers, etc.

In the process of further analysis of the current business situation and its development, it turned out that a large component of the cost of production is occupied by costs that are not directly related to the process and volume of production. Due to the constantly growing competition, the end consumers of the products are becoming more and more selective, the costs of advertising and marketing are increasing noticeably, and the life cycle of the products is decreasing. All this requires revision of views on planning of commercial activity. Based on these prerequisites, a new concept of corporate planning, called MRP II, was born.

4.1. Basic modules of the MRP 11 system

In any production enterprise there is a set of standard principles of planning, control and management of functional elements. These elements are production shops, functional departments, management apparatus, etc. For complex automation of large enterprises MRP II-systems are used, which include the following functional modules [4]:

1. Business development planning;
2. Sales planning;
3. Planning requirements for raw materials and supplies;
4. Production planning;
5. Production capacity planning;
6. Fulfillment of the production plan;
7. Fulfillment of the plan of material requirements;
8. The implementation of feedback.

Let's consider the purpose of each module in more detail. The ***business development planning*** module determines the purpose of the company (its mission): its market niche, evaluation and determination of profits, financial resources, i.e. determines in conditional financial units, production and sales volumes and assesses what amount of funds

It is necessary to invest in the design and development of the product in order to reach the planned level of profit. The output of this module is the business plan.

The ***sales planning*** module estimates (in units of finished goods) the volume and dynamics of sales needed to fulfill the developed business plan. In this case, changes in the sales plan entail changes in the results of other modules.

The ***raw material requirements planning*** module based on the production program for each type of finished product generates the required amount of materials and schedules the purchase and/or internal production of all components of this product and, accordingly, their assembly.

The ***production planning*** module approves the production plan for all types of finished products and their characteristics. Each type of product within the product line has its own production program. Thus, the totality of production programs for all types of manufactured products represents the production plan of the enterprise as a whole.

The ***capacity planning*** module converts the production plan into final units of capacity utilization (machines, workers, laboratories, etc.).

Modules responsible for the ***implementation of production plans*** and ***material requirements,*** are used to monitor and report on the activities of the enterprise.

The ***feedback*** module allows you to discuss and resolve problems with component suppliers, dealers, and partners. Feedback is especially necessary when changing individual plans that have proven to be unrealistic and are subject to revision.

4.2. Algorithm of MRP 11 system operation

A schematic representation of the MRP 11-system algorithm is shown in Fig. 7. 7.

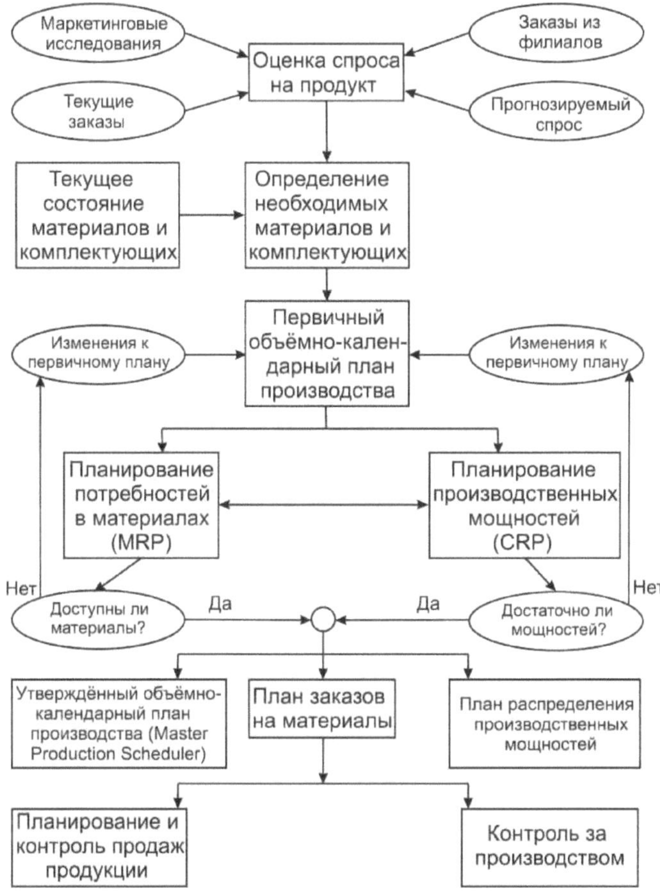

Fig. 7 Algorithm of MRP 11 system operation

The following steps are performed [1,4]:

1. Gathering and analyzing information about the demand for a certain end product.

2. Based on the information about the status of components, determine what

materials (components) are available.

3. The primary production plan (IPPP) is drawn up, on the basis of which the production capacity is checked and the planning of material requirements is carried out.

4. If the available resources are sufficient, the primary plan is taken as a basic volume-calendar production plan, the formation of a plan of orders for materials with the timing of deliveries and a plan for allocation of production capacity to determine the optimal load.

5. There is control over the production of the final product and its further sale.

The advantages gained from the implementation and use of the MRP II standard EIS are in the possibility of using the following functional elements:

- Obtaining operational information about the current results of the enterprise as a whole, and with full details of individual orders, types of materials, the implementation of plans;
- Long-term, operational and detailed planning of the enterprise's activities with the possibility of correcting the planned data on the basis of operational information;
- Solving the problems of optimization of production and material flows;
- Actual reduction of inventories in warehouses;
- Planning and control of the entire production cycle with

the ability to influence it in order to achieve optimum efficiency in the use of production facilities, all types of resources and customer satisfaction;

- Automation of the customer service department with full control over payments, product shipment and due dates;
- A financial reflection of the enterprise as a whole;
- Significant reduction in non-production costs;
- Protection of investments made in information technology;

• The possibility of phased implementation of the system, taking into account the investment policy of a particular enterprise.

4.3. Hierarchical organization of plans in the MRP 11 system

The MRP II system is based on the hierarchical organization of plans, shown in Fig. 8 [1-2, 4]. The lower level plans depend on the higher level plans, i.e., the higher level plan provides inputs, targets, and/or some kind of constraints for the lower level plans. In addition, these plans are linked in such a way that the results of the lower level plans have an opposite effect on the higher level plans.

If the results of a plan are not feasible, then that plan or higher level plans must be revised. That is, it is possible to coordinate the supply and demand of resources at a particular planning level and resources at higher planning levels.

Consider the main levels of enterprise planning:

• *Strategic planning* is long-term planning, usually for one to five years, based on macroeconomic indicators such as economic trends, technological change, market conditions, and competition. Strategic planning usually extends to each year of the five-year period and presents top-level planning figures (goals).

• *Business planning* is usually annual planning, which is also done on an annual basis. It is often revised during the year. It is usually the result of a meeting of the management team, at which plans for sales, investments, development of fixed assets, capital needs, and budgeting are summarized. This information is presented in monetary terms. The business plan sets sales and production targets and other lower-level plans.

• *Planning by nomenclature groups is a* plan of sales and production volumes, divided into 10-15 assortment groups. The result is a production plan that is revised monthly, taking into account the previous month's plan, actual results, and business plan data.

From the description it is clear that the most common (often performed) is planning by nomenclature groups, the result of which is the development of a plan for sales and production, which includes the following elements:

1. *Sales volume;*
2. *Production;*
3. *Stocks;*
4. *Incomplete production;*
5. *Shipping.*

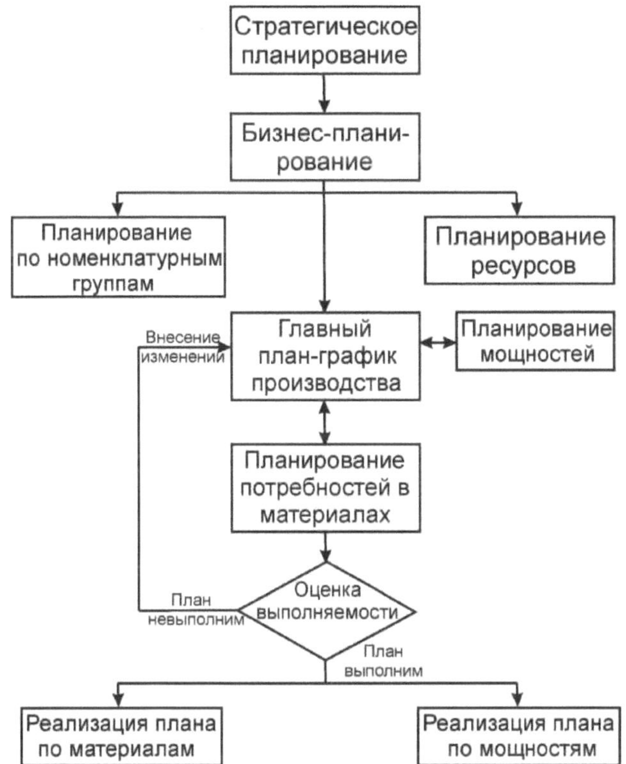

Fig. 8 Hierarchical organization of MRP 11-system plans

Let's consider the purpose of each element. ***Sales volume*** and ***shipment*** are forecasted values, because they depend on external data, which are not directly controllable. ***Production volume*** is planned, because it is an internal indicator that can be directly controlled. ***Inventory*** and ***work-in-progress plans*** are controlled indirectly by manipulating data from ***sales volume*** forecasts, ***shipment volume forecasts***, and/or ***production volume plans***.

Inventory levels and ***work-in-progress*** are managed differently, depending on the types of products the company produces or sells. For example, the ***planned inventory volume*** is

is an important factor, especially for those companies that produce to stock. ***The work-in-progress plan*** is an important factor for those companies that produce to order.

The basic element *of sales* and ***production planning*** is ***the production plan***. Despite its name, it is not just a production plan. It requires the availability of the necessary amount of resources throughout the company as a whole. If the marketing department plans to increase sales of a particular product range, engineers must ensure that the necessary equipment is available; the supply department must ensure that additional supplies of materials are available; the human resources department must ensure that additional labor is available and that new shifts are organized. In addition, the required amount of capital (to pay for the additional resources and inventory) must be available.

The production plan will not be fulfilled if the necessary amount of required resources is not available. ***Resource planning*** is long-term planning that estimates the amount of key resources, such as personnel, equipment, buildings, and facilities that are necessary (to execute the production plan) and available. If additional resources are needed, the business plan will need to be revised.

Resource planning affects only key resources and is made for the duration of the production plan (usually one year). A resource is considered key if its value is relatively high, or if its delivery time is long, or if other resources depend on it. Resources can be ***external*** (supplier capabilities) or ***internal*** (equipment, storage space, money).

Having planned your resources, you must develop ***a master production schedule (MPS)***, which is a production plan overlaid on a timeline, based on the production plan. The CMP shows what will be produced, when, and in what quantities. Due to the fact that the production plan is expressed in units such as rubles, hours, and tons, it is necessary to make some steps to transform it in order to obtain the CMP. The planned volume indicators for the assortment group should be translated into planned volumes and terms for each product of this group separately. Depending on the type and volume of output products, the GPGP can be broken down into weekly, daily, and even shift plans.

One of the main goals of the GPGP is to provide a buffer. GPGP distinguishes sales department forecasts and needs from MRP (material requirements planning). The approach is as follows: forecasts and sales orders (customer orders) express demand (or shipment), while the GPGP reflects what will actually be produced according to the available demand. According to the CMP, it is possible to produce products at a time when demand is low, and vice versa. This is the case in the production of products for which demand is seasonal.

The main problem in making the GPG is to determine which products (components) should be planned by the planning department, and which should be automatically controlled (MRP system). The products planned by the planning department are those products, the planning of which must be controlled by humans. Products planned by the MRP system, i.e. automatically, do not require this degree of control (they only depend on the GPGP). The definition of how the planning of the individual product depends on the product types and processes. Usually a small number of products have to be controlled by the planning department.

Like resource planning, *overall capacity planning* is long-term and is performed on key resources. This process uses GIP data, not production plan data. For example, if the GPGP is expressed in terms of volume and time, then general capacity planning is used to create a more detailed plan, which is useful in assessing the average needs of the company as a whole, as well as for evaluating the GPGP.

Historically, MRP was designed to control and replenish inventory. Under MRP II, its use has been extended to capacity requirement planning, prioritizing, and closing the entire planning chain.

Resource Planning and *General Capacity Planning (CRP)* is a top-level planning used to manage resources such as physical equipment. *Capacity requirement planning* is more detailed planning. Job utilization is calculated based on the *process route* for manufacturing a product, which determines exactly how a given type of product is produced. *A process route* is similar to an instruction manual - a set of steps or process operations that must be performed to make a certain product (or product). Each process operation is performed at a specific workplace, which may consist of one or more people and/or equipment.

As materials move from supplier to customer, they move through *the supply chain* (or market channel). If we visualize this graphically (Fig. 9), the supply chain looks like supply and demand flows between suppliers and customer company divisions, between these divisions and customers, or between different divisions of the same company. DRP-system (Distribution Requirement Planning) coordinates demand, supply and resources between divisions of one or more companies.

Supply flow

Seller Supply Chain Buyer

Demand stream

Fig. 9 Generalized graphical representation of the supply chain

There may be two or more levels of production and/or distribution units in the supply chain. These units may be dependent on each other in various ways. The important thing is that one division can supply products to another division. For example, some company manufactures goods in one division and sells them from a separate sales warehouse. Such a supply chain is illustrated in Figure 10.

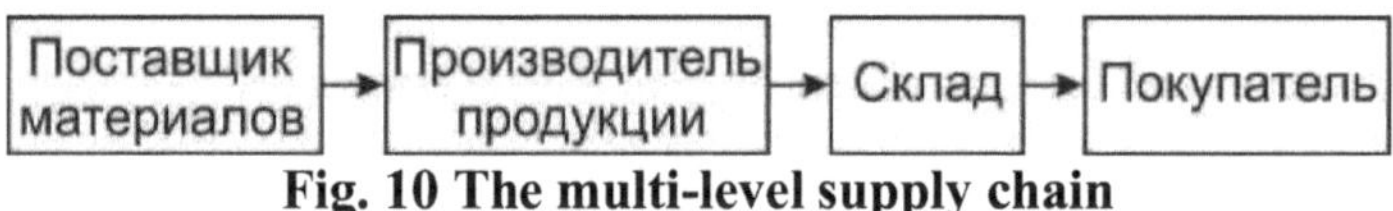

Fig. 10 The multi-level supply chain

If the company has a single (main) distribution center, which supplies products to warehouses of regional branches, the supply chain will look as shown in Figure 11.

If a company has production facilities (plants, workshops) in two different locations, it is possible to build a distributed supply chain, an example of which is shown in Figure 12.

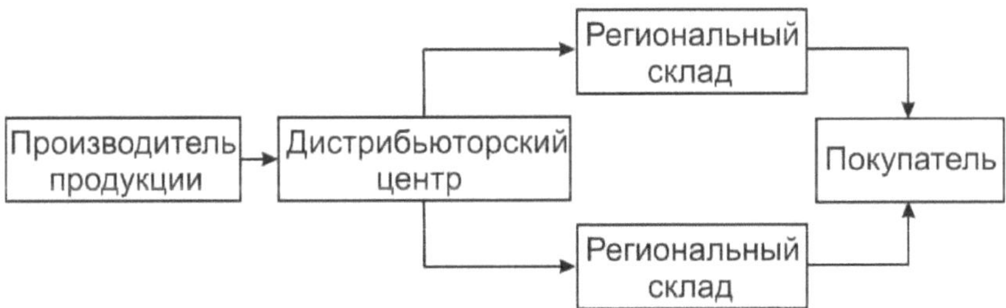

Fig. 11 Supply chain with a single distribution center

Fig. 12 Distributed supply chain

In ***planning the supply and demand for materials*** between units, the following three main tasks must be accomplished:

1. Timely receipt of necessary materials from other units;

2. Timely delivery of components issued to other units;

3. Obtaining components that are currently available (in stock).

All three tasks are solved similarly to the principles of solving similar problems in the MRP-system, but in the presence of DRP-system is possible to control supply and demand in the context of the various structural units.

4.4 Role of feedback in the MRP 11 system

The feedback function plays a key role in the MRP II system.

For example, if suppliers are unable to deliver materials and components within the agreed timeframe, they should send a delay report as soon as they become aware of the problem [1, 4]. In plants managed by MRP II class systems, delivery dates are as close as possible to the time of actual need for supplied materials.

Therefore, it is extremely important to notify the system in advance of possible problems with orders. In this case, the system should generate a new production capacity work plan, according to the new order plan. In some cases, when delayed orders are far from being an exception, the MRP II system sets the volume of the minimum maintenance of the stock of "unreliable" materials in stock (safety stock).

MRP II systems are firmly in the life of large and medium-sized production organizations. The main and effective feature of these systems is the ability to plan company needs for short time intervals (weeks or even days) and to provide feedback (e.g., automatically change previously made production plans in case of supply failures or equipment breakdowns) by entering real-time data on problems into the system.

The operating algorithm of the MRP II system is aimed at the internal modeling of the whole business area of the company. Its main purpose is to take into account and analyze by computer all intra-commercial and intra-production events: all those that occur at the moment and all those that are planned for the future.

Any MRP II system has certain tools for conducting planning. Listed below are the system methodologies that are the fundamental control levers of any MRP II system:

1. Methodology for calculating and recalculating MRP and CRP plans.

2. The principle of storing data on intra-production and intra-commercial events, which are necessary for planning.

3. Methodology for describing working and non-working days for resource planning.

4. Setting the planning horizon.

These methodologies and principles are not universal and are determined based on the formulation of a specific problem, in relation to the production enterprise.

5. ERP standard

ERP (Enterprise Resource Planning System) systems are designed to manage the financial and economic activities of enterprises [3, 5-6]. This is the "top level" in the hierarchy of enterprise management systems, affecting the key aspects of its production and commercial activities. Such systems are created to provide management with information to make management decisions, as well as to create an infrastructure for electronic data exchange of the enterprise with suppliers and consumers.

All enterprises are unique in their financial and economic activities. However, progress in the development of software solutions for ERP tasks is due to the fact that along with the specifics, it is possible to identify tasks that are common to enterprises of different types of activities (various industries, services, telecommunications, banks, government agencies, etc.). Such common tasks include management of material and financial resources, procurement, sales, customer orders and supplies, personnel management, fixed assets, warehouses, business planning and accounting, bookkeeping, settlements with customers and suppliers, maintenance of bank accounts, etc.

5.1. The need to switch from MRP II to ERP

Over time, the MRP II system has evolved into Enterprise Resource Planning (ERP), sometimes also called enterprise-wide resource planning. ERP is based on the principle of creating a single data repository containing all the business information accumulated by an organization in the course of its business operations, including financial information, production-related data, personnel management, or any other information.

This approach eliminates the need to transfer data between individual systems. In addition, any part of the information that the organization has becomes available at the same time to all employees with the appropriate credentials. Once the ERP concept was proven to be feasible in a manufacturing environment, it became possible to create a single information resource used by the commercial organization as a whole.

It is very rare that organizations try to implement ERP-systems projects on their own. It is equally rare for a software vendor to be involved in this process. To implement such software, companies hire contractors, who are consultants experienced in working with ERP-class software directly in the client's field of professional activity.

ERP systems, in contrast to MRP II, are focused on the management of the "virtual enterprise", which reflects the interaction of production, suppliers, partners and consumers, which are autonomously operating enterprises or corporations.

The ERP adds controls for multinational corporations, including support for multiple time zones, languages, currencies, accounting and reporting systems. These differences are less about logic and functionality and more about infrastructure (Internet/Intranet) and scalability - up to several thousand users. Requirements for flexibility, reliability and performance of software and computing platforms are constantly increasing.

Requirements for the integration of ERP systems with applications already in use at the enterprise (e.g., design, production preparation, production flow and process control, billing and customer billing systems, etc.) as well as with new developments are increasing. ERP system can not solve all the problems of industrial enterprise management and is often seen as a basis for integration with other applications. New ERP systems pay a lot of attention to decision support tools and tools for integration with data warehouses (sometimes included in the system as a new module).

5.2. Functional modules of ERP systems

ERP-systems implement the following basic functional blocks (modules) presented in Fig. 13 [1, 3, 5-6]:

- *Sales and production planning.* The result of the block action is the development of the production plan for the main types of products.
- *Demand management.* This block is designed to forecast future product demand, determine the volume of orders that can be offered to the customer at a particular time, determine distributor demand, demand within the enterprise, etc.
- *Aggregate capacity planning.* It is used to specify production plans and determine the degree of their feasibility.
- *The main production plan (production schedule).*

Products are defined in final units (products) with lead times and quantities.
- *Planning of material requirements.* Types of material resources (prefabricated assemblies, finished units, purchased products, raw materials, semi-finished products, etc.) and specific terms of their delivery to fulfill the plan are determined.
- *Product Specification.* It defines the composition of the final product, material resources required for its production, etc. In fact, the specification is a link between the main production plan and the plan of material requirements.
- *Planning for capacity requirements.* At this stage of planning, production capacity is defined in more detail than at the previous levels.
- *Routing/work centers.* With the help of this block, both the production capacities of the different levels and the routes according to which the products are produced are specified.
- *Checking and adjusting shop floor capacity plans.*

- *Management of procurement, inventory, sales.*

- *Financial management* (general ledger maintenance, settlements with debtors and creditors, fixed assets accounting, cash management, financial planning, etc.).).

- *Cost management* (accounting of all enterprise costs and costing of finished products or services).

- *Project/Program Management.*

Fig. 13 Functional blocks of the ERP system

In accordance with modern requirements ERP-system must include the following additional modules (functional blocks) in addition to the core that implements the MRP II standard (or its analog for continuous production):

- Supply Chain Management (SCM), formerly DRP (Distribution Resource Planning).

- Advanced Planning and Scheduling (APS).

- CRM (Customer Relationship Management).

- EU Electronic Commerce.

- Product Data Management PDM (Product Data Management).

- Decision Support Systems (DSS) module.

- Stand-alone module responsible for system configuration

SACE (Stand Alone Configuration Engine).

- The FRP (Finite Resource Planning) is a detailed resource planning process.

ERP systems have developed advanced tools for adjustment (configuration) and adaptation, including those that are applied dynamically during the operation of the systems.

The use of ERP methodology has now become ubiquitous. Manufacturers who plan to succeed in an increasingly competitive market must actively use ERP systems in order to match the production efficiency of their competitors.

ERP systems have developed advanced tools for adjustment (configuration) and adaptation, including those that are applied dynamically during the operation of the systems.

5.3 Stages and principles of ERP-systems implementation

The process of implementing the ERP-system at a particular enterprise consists of the following steps (stages) [1,3, 5-6]:

- Development of automation principles and strategy.

- Analysis of the business activities of the enterprise and the nomenclature of products.

- Reorganization of business processes in accordance with the ERP standard.

- Choice of CIS for the tasks of the enterprise.

- Implementation of the chosen system.

- Operation of the software product and user support.

Let's consider each of the stages in more detail. ***The stage of developing the principles and strategy of automation*** involves the definition of basic principles used in the automation of a particular enterprise, which largely depend on the range of products manufactured. These principles can include goals and objectives of automation, in which the areas of the company and the sequence in which they will be automated, automation methods (by area, direction, integrated automation). Various kinds of constraints, such as financial or time constraints, are taken into account.

A strategic automation plan should take into account the following factors:

- the average period between changes in the main production technology
- the average lifetime of the products produced by the company and its modifications
- announced long-term plans of technical solution providers in terms of their development
- the amortization period of the systems used

- Strategic plan for enterprise development, including plans for mergers and splits, changes in headcount and product mix
- Planned changes in personnel functions

Automation is one of the ways to achieve strategic business goals, not a process developing according to its own internal laws. Automation strategy should be driven by the business strategy of the company: the company's mission, direction, and business model. Therefore, automation strategy is a plan aligned in terms of timing and goals with the overall strategy of the organization.

The second important feature is the degree of consistency between the priorities of automation and business strategy, namely what goals are to be achieved: reducing the cost of production, increasing the number or assortment, reducing the cycle: the development of new products and services, etc.

Typical problems that arise when developing an automation strategy, as a rule, are associated with such factors as the state of the information technology market, determining the effectiveness of investment in information technology, the need to restructure of an enterprise activity in the implementation of information technology and many others.

Once the automation strategy has been selected, you can move on to the next step of ERP system implementation. The ***analysis of business activities of the enterprise and the range of products manufactured*** means the collection, analysis and presentation of information about the activities of the enterprise in a formalized structured form suitable for the selection and further implementation of an automated system. Depending on the chosen strategy for automating the enterprise technologies of collecting and presenting data may be different. The final presentation of information at the activity analysis stage plays a key role in all further work and in the final success of automation and software product implementation.

Re-engineering business processes according to the ERP standard involves changing some of the operations performed so that they are easier to automate. The reason is that in the process of development each enterprise modifies its key processes in order to perform certain functions. As a result, unnecessary operations appear which could have been avoided (without losing the quality of the final product). It is after a detailed analysis of all business processes it is possible to identify and sift out all unnecessary operations, i.e. to reorganize business processes.

The choice of a CIS for the tasks of the enterprise involves the analysis of the software products available on the market that are optimally suited for the object of automation. This is a multi-criteria task, and setting objective criteria for the selection of a particular system is directly related to the quality and completeness of the previous stages of the selection chain. Virtually all objective considerations that guide the choice of the system (functionality, system cost and total cost of ownership, development prospects, support and integration, technical characteristics of the system, etc.) are determined in the previous stages. When all of the previous steps are thoroughly considered, system selection is no longer a problem.

The implementation of the selected corporate information system is the most serious and responsible stage in the whole process of automation. It is what determines whether all the investments made by management will pay off. There are the following main strategies for implementing the system:

- Parallel strategy - when the old and new systems work simultaneously, and their output documents (reports, graphs) are compared. If they agree for a long time, the transition is made to the new system.
- "Leap." This strategy is attractive, but not recommended, because it implies complete and lightning-fast automation of all key business processes of the enterprise. Such a strategy can give a positive result only at a small enterprise, and for large corporations it is doomed to failure.
- "Pilot Project." This is the most commonly used strategy,

assuming complex automation of a small (limited) number of key processes. The scope of the strategy is a small area of activity. This approach reduces risk and costs, so it is the most reliable. Almost all enterprises use such tactics nowadays.

- "Bottleneck. With this approach, the implementation plan is performed only for the "bottleneck" (a certain area of automation) and for the people working in it. Data accuracy is improved only for products in this "bottleneck"; retraining - only for people working in it; cost-benefit analysis is done only for it, etc.

The stage of software product operation and user support is the final step and therefore confirms the correctness of all previous stages. This is the most costly stage, because the modernization of software and hardware, caused by the physical and moral aging of the ACS components, the need to track changes in legislation; the need to refine the system to meet new requirements of its users, ensuring information security in the process of operation require huge capital investments.

The cost of operating the system within the enterprise can and should be reduced by the qualitative study of the preceding stages, mainly by developing a strategy for automation and the implementation of the choice of system.

5.4. Main advantages and disadvantages of ERP systems

The benefits of successful ERP implementation include

The system at a particular enterprise can be attributed to:

- Reducing the cost of products and services through the efficiency of operations;
- Reducing the time to market for products;
- Reducing costs and defects in production;
- Improving the quality of products;
- Closed loop order processing.

The disadvantages of existing ERP systems are:

- Inner focus;
- Limitation of functions to production only

administration;

- Lack of sales, marketing and new product development functions;
- The system's response to market changes is delayed;
- The principles and efficiency of operations can be borrowed and improved by competitors.

The ERP concept has become popular in the manufacturing sector because resource planning has reduced lead times, reduced inventory levels in warehouses, and improved customer feedback while reducing the administrative burden.

To optimize supply chain management, the concept of Supply Chain Management (SCM) was created, which is supported by most MRP II class systems. SCM, acting as a component of a company's overall business strategy, makes it possible to significantly reduce transportation and operating costs by optimal structuring logistics schemes supplies. Implementation of this functionality allows to optimize procurement and sales processes. SCM provides the ability to automatically import and store price lists of suppliers and, if necessary, competitors. Based on the data from the price lists of suppliers information about new products is formed, which allows to choose the best assortment. Analytical information about competitors' prices allows to optimize sales process, offer competitive prices and thus increase company profits.

Orders for delivery of goods are formed in the names of commodity items from the suppliers' price lists. To optimize debt supply and to maintain the assortment, the expert system offers the best purchasing options, providing information on stale goods, or warns about the imminent expiration of the most demanded items by the buyer. When placing an order, the system determines the best price for purchasing materials and selling finished goods for each item, based on prices from suppliers' and competitors' price lists, the cost of delivery from the supplier's warehouse, and other parameters. The concept of customer relationship management (CRM) (Customer Relations Management) is also widespread [7-8].

5.5. *The basic concepts of CRM-strategy*

In the conceptual construction of traditional ERP enterprise resource management systems, the customer is viewed as an element of the outside world, not integrated into the business processes served by the ERP system. The meaning of this arrangement of systems to manage the company was determined by the strategic focus of business to optimize only the internal activities of the enterprise, which is now obsolete. Many divisions of the enterprise, working with the outside world, separated from each other, although interacting with the same counterparties. The lack of a unified approach in working with the client has a negative impact on the effectiveness of work in the market - the company loses the opportunity to increase sales and increase customer loyalty.

Modern marketing research has shown that having a solid base of loyal customers is the main and perhaps the only factor in the sustainability and prosperity of a business. To integrate the customer inside the company, to give him a personalized service - the main task of business planning. As part of this task was born a whole strategy aimed at shifting the concentration of efforts to bring order within the company towards customer service - called CRM.

CRM is a company strategy that defines interaction with customers in all organizational aspects: it deals with advertising, sales, delivery and customer service, design and production of new products, billing, etc.

The strategy is based on the following conditions [7-8]:

• ***The presence of a single repository of information*** containing information about all cases of interaction with customers.

• ***Synchronized management of multiple interaction channels*** (i.e., there are organizational procedures that regulate the use of this system and information in each division of the company).

• ***Continuous analysis of collected information about customers*** and making appropriate organizational decisions, such as ranking customers based on their importance to the company, the development of an individual approach to customers according to their specific needs and requests.

There is the following classification of CRM-systems in three key areas [1,7]:

1. ***Operational CRM*** includes applications that provide quick access to information on a particular customer in the process of interacting with him within the framework of normal business processes - sales, shipping, service, etc. It requires tight integration of different systems, clear organizational coordination of the process of interaction with the customer through all channels. At the moment, the vast majority of software complexes accounting interactions with customers are operational CRM.

2. ***Analytical CRM*** involves the synchronization of disparate data sets and

the search for statistical patterns in these arrays to develop the most effective marketing, sales, customer service strategy, etc. Requires a good integration of systems, a large amount of accumulated statistical data, effective analytical tools. Analytical CRM is less popular than operational CRM and is closely related to the concepts of Data Warehousing (data warehouses), and Data Mining (data analysis).

3. ***Collaborative CRM*** gives the customer much more influence over the design, production, delivery and service processes of the final product. Requires technologies that allow cost-effectively connecting the customer to the collaboration within the company's internal processes. Examples of features that are implemented in

Collaborative CRM system are:

- Collection of customer suggestions in product design.
- Customer access to experienced product samples and the opportunity for feedback.
- Reverse pricing, in which the customer describes the requirements for the product and determines how much he is willing to pay for it, and the manufacturer responds to these suggestions.

The latter case is the rarest aspect of the use of CRM, requiring for its implementation a radical restructuring of internal organizational mechanisms. But those few companies that implement it have already achieved unprecedented rates of return on investment.

There are almost no systems supporting collaborative CRM on the market, also because the collaborative process in most cases is purely individual and needs to be automated through an extremely flexible CRM system. In addition, this system should be based on the cheapest and most open technologies (for example, Internet technologies) to reduce the cost of building an interface between the company and its customers.

6. CSRP standard

At the end of the last century, it became clear that ERP systems were simply necessary in enterprises, but their management capabilities in today's fast-paced business world were clearly insufficient. At the beginning of the new century, more powerful production management tools emerged that built on the solid foundation of ERP and focused on integration with customers. This decade's production planning system has two focuses - on production efficiency and on creating customer value. This new paradigm - strategy is called Customer Synchronized Resource *Planning* (*CSRP*) [9].

6.1. Principles of order formation and processing in CSRP-systems

CSRP leverages proven, integrated ERP functionality and redirects production planning from production down to the customer. CSRP provides actionable methods and applications for creating value-added products for the customer.

The implementation of CSRP requires [1,9]:

• . *Optimize production activities* (operations) by building an effective production infrastructure based on ERP methodology and tools.

• . *Integrate the buyer and buyer-focused divisions* of the organization, with major planning and

production units.

• . *Implement open technologies* to create a technology infrastructure that can support the integration of buyers, suppliers and production management applications.

Synchronizing the activities of the organization's customer-facing and buyer-facing departments with the company's executive and planning center provides the ability to identify favorable opportunities to create differences that support competition. The rise of manufacturing, by injecting real-time customer requirements into the organization's daily planning and production systems, forces company executives to expand their focus beyond the manufacturing process and consider critical product and market factors.

In the standard under consideration, order processing expands and, instead of a simple order entry function, integrates sales and marketing functions with the customer and proceeds as follows:

• ***Vendors no longer place orders***. They work with the buyer and at his workplace to form orders, identifying the buyer's needs, which are dynamically translated into product requirements and production. Order configuration technology makes it possible to check the feasibility of an order before it is placed.

• ***Order processing is expanding to include prospect information***. Leading contact management systems integrate with the order creation and production planning process to provide information on required resources before an order is placed. Market trends, product demand and information about competitors' offerings are linked to key business processes.

• ***Static pricing models are replaced by a pricing tool*** that allows you to determine the value of each product for each customer, if necessary. Accuracy in cost and profit calculations for product sales is increased. CSRP redefines customer service and extends it beyond the usual phone support and issuance of account statements. With the CSRP model, customer service becomes the command center for the organization. The customer support center is responsible for communicating critical customer information to the organization's executive centers. The following applications and technologies are implemented and used:

• ***User support applications are integrated with key planning, production*** *and management applications*. Critical customer and product information is delivered in advance to production, sales, research and development, and other departments.

• *Web-based technologies enhance customer support, including remote, 24/7, self-customizing support*. Key performance systems automatically change, increasing the ability to deliver products and services faster to customers.

• *Customer support centers become sales and user support centers*. Integration with sales, order processing and management provides the knowledge and infrastructure to turn customer support into sales activities, providing a channel for promoting new
and related products and services.

The planning of production and all activities is redefined and becomes the planning of customer orders and the dynamics of production. This is realized by means of the following technologies:

• *Direct integration with order configuration information* allows manufacturing departments to increase the integrity of the planning process by reducing repetitive work and reducing interruptions due to order overflow. Improved production scheduling enables manufacturers to provide better estimates of delivery times and improve on-time delivery.

• *Production planning now makes it possible to optimize operations based on actual customer orders* rather than on forecasts or estimates. With real-time access to accurate customer order information, scheduling departments can dynamically change work grouping, customer order sequencing, acquisitions and subcontracting to improve customer service and reduce costs.

• *Buyer requirements for a product can be transferred directly from the buyer to the subcontractor or supplier*, eliminating the errors and delays encountered when translating buyer orders into material purchase orders. Changes to a buyer's order can result in automatic changes to component suppliers' orders, reducing rework and delays. Product quality and the correct ordering of core materials can be significantly improved, and delivery cycles can be reduced.

6.2. Main advantages of CSRP systems

With the CSRP business model, traditional business processes are redesigned to serve customers and create products that meet their needs. Implementing CSRP applications pushes enterprise leaders to change. The internal focus of traditional manufacturing structures, segmented by department and functionality, is refocused outward. CSRP allows for a bi-directional free flow of information between the customer and the manufacturer.

The considered concept of building information systems allows you to implement a more accurate management of production schedules in conditions of limited capacity (the so-called APS-task - Advanced planning and scheduling - advanced management of production schedules).

APS-type systems make it possible to solve such tasks as urgent order execution in the production schedule, assignment of tasks taking into account priorities and restrictions, and rescheduling using a full-featured graphical interface. Thanks to a fundamentally new mathematical apparatus used in CSRP systems, the calculation of typical MRP tasks is much faster than before. A typical example of situation when APS-systems application is effective is additional urgent order at the enterprise, where production program has already been formed and is being executed. While attractive, the execution of a new order can have serious consequences: delayed fulfillment of previously accepted orders, disruption of production cycles and, ultimately, financial losses for the enterprise. In this case, it is necessary to decide whether to accept the order, and if so, what should be its cost to the buyer.

The implementation of the CSRP concept in a particular enterprise allows you to manage customer orders and all work with them by an order of magnitude more accurately than in previous systems. It is now possible to change the production schedule on an hourly basis - with each new order, it is possible to completely recalculate the production program taking into account the priority strategies of the enterprise. With the classic ERP-system, it was practically an impossible task. The calculation of the detailed cost of an order or its individual components is now possible already at the stage of its registration. Variations in product specification or process chain that are often required in printing and other industries can be taken into account. All additional operations for testing and order administration, not to mention after-sales service, are also taken into account.

In the framework of CSRP and similar methodologies it is very important to integrate with the enterprise resource management system software products of third firms that implement the specific tasks of management and calculation of resources, specific to a particular enterprise. In general, the application of new methodologies of enterprise resource management allows the company to feel confident even in an unstable market and rapidly changing macroeconomic environment. CSRP forces a reconsideration of all business practices, focusing on market activity rather than on production activities. Business processes are synchronized with customer activities. For an example, consider the order processing process. Customer-Synchronized Resource Planning (CSRP) offers a business model and set of tools that can make partnering with the customer possible and habitual.

Customer Synchronized Resource Planning (CSRP) offers a new set of business rules that make it possible for a manufacturer to create customer value - to develop solutions and services that will make them (the manufacturers) necessary for customers. The business functions offered by CSRP systems allow a manufacturer to personalize products and develop solutions and services that will make them necessary for customers. Increasingly, competitive advantage is defined as the ability of manufacturers to meet the unique needs of each individual customer at all times and to do so consistently.

7. ERP II standard

The authoritative consulting company Gartner Group declared the end of the era of ERP systems in 1999. It was replaced by ***ERP II*** (Enterprise Resource and Relationship Processing) [1, 10], ***a concept for managing internal resources and external relationships of an enterprise.*** The main idea of ERP II-system is to go beyond the tasks of optimization of internal processes of the organization and optimization.

At present, ERP II systems are the pinnacle of progress in the field of information management solutions. Historically, the development of new concepts has proceeded along the lines of absorption of tried and tested standards and the formation of new, proven solutions on their basis (Fig. 14).

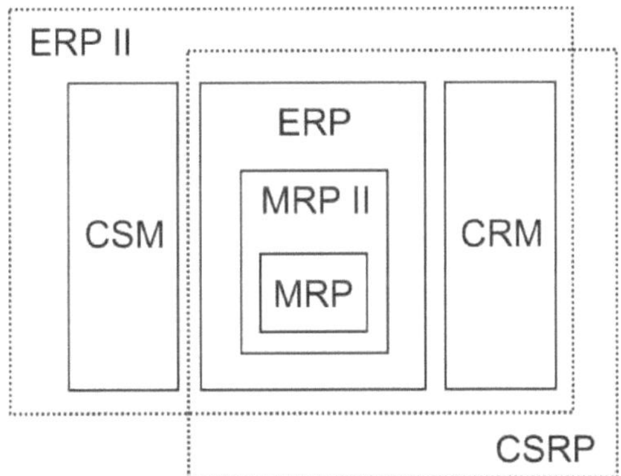

Fig. 14 Standards of enterprise management

According to the ERP II concept, further development of corporate information systems for business management will follow the following directions. First, the key focus will be on deepening rather than expanding existing functionality, i.e., supporting as many execution options as possible

(implementation) of typical business tasks. Secondly, creation of specialized industry-specific solutions that are closely integrated with basic ERP functionality will become much easier. Thirdly, tools to manage inter-corporate business processes will play an increasing role. All highlighted directions of development should be considered as a whole and advancement in each of them by itself does not guarantee that the system belongs to the ERP II class.

The evolution of automation systems has had a direct impact on business practices and technologies. In the search for competitive advantages and the desire to provide a higher level of service, market leaders are carefully analyzing the potential of all the latest solutions and technologies. Not long ago, e-commerce was considered a rarity and was only available to large companies producing huge batches of products. Today, thanks to the ubiquity of the Internet, it is impossible to imagine even a relatively small store without e-commerce. Interactive interaction between companies and their regular partners via the Internet is on the rise. This means much greater access to corporate information for external users, and should therefore be based on the principles of security and trust in the partner as well as on agreed rules of operation.

The ERP II concept involves the development of basic ERP functionality by upgrading existing modules, i.e. the systems from version to version have more new additional features in the financial, logistics and production loops, implemented separate accounting for different legal entities within a single application with a common database.

7.1. Differences between ERPII and ERP systems

Gartner Group defines ERP II as a business strategy and a set of industry-specific applications that enable and optimize internal and external business processes, joint operational and financial initiatives. The key financial areas of ERP II are accounting, buying and selling products and materials, order entry and costing.

The purpose of ERP II is not only to optimize the resources and transaction processing of traditional ERP systems, but also to use information. ERP incorporates these functions in the process of business-to-business cooperation.

The role of ERP is thus not limited to e-commerce sales and purchasing. The scope of ERP II extends beyond ERP to the non-manufacturing industry.

The internal functions of these industries go beyond a broad understanding of production, distribution, and finance and integrate actions specific to an industry sector or a particular industry. Concentrated on the Internet, designed to integrate ERP II product architecture, they are so different from monolithic ERP architectures that they require a complete transformation. ERP II extends ERP's ability to store all data within the enterprise to the ability to handle data distributed across the trading community. Figure 15 shows the main differences between ERP II and ERP [Yu].

Thus, ERP II is the result of the development of ERP methodology and technology towards a closer interaction enterprise with its customers and counterparties. In this case, the management information of the company is not only used for internal purposes, but also serves to develop cooperative relationships with other

organizations.

Parameter	ERP system	ERP II system
The role of the system in the enterprise	Process optimization	Participation in the chain , that ensures an increase in profits , Creating conditions for co-commerce
Area of application	Production and distribution	All segments and sectors
Functions	Production, trade (distribution) and financial processes	Interdisciplinary and industry sectors , specific production processes
Types processes to be automated	Internal, hidden	Connected at the external level
Architecture	Monolithic, closed, usually based on Web-technology	Open, component, Internet-oriented
Data	Generated and used internally	Designed for both indoor and outdoor use

In addition to the new managerial orientation of ERP II systems are also characterized by some technological features. This refers to the Internet-oriented architecture, which differs significantly from the architecture of traditional ERP-systems. This is due to the fact that managerial information previously stored and used only within the enterprise, now should be available for information systems of customers and partners. Thus, traditional client-server architecture begins to give way to Web-clients and distributed component technologies.

7.2. Problems of implementing ERP II-systems

The key technologies in the ERP II concept are "collaboration" and the Internet. Today's largest ERP vendors - J. D. Edwards, Oracle, PeopleSoft and SAP - are making their way to ERP II. However, this process of "moving to ERP II" threatens to drag on for the following main reasons [10]:

• ***The collaborative strategies of these companies*** are generally aimed correctly, though there are few examples of actual collaborations and they take asynchronous (e.g., collaborative forecasting) rather than synchronous form (e.g., the available supply chain).

• ***All four suppliers have vertical initiatives***, but most of them are too broad and marketing-oriented or too optimistic.

• ***Functions of "add-ons", such as***   ***control customer relations or supply chain management*** used by four suppliers are not vertically focused and are mostly inferior to advanced suppliers to the point that most businesses are forced to make difficult strategic choices between a single or advanced supplier.

• ***Most major ERP vendors have marketplace initiatives designed to support inter-enterprise buying and selling***, but they provide little or no workflow capability between companies in their core products.

• ***Vendors have made some progress in architecting their own products***, but none of them has much in the way of object-oriented products designed for integration. Moreover, none of them have ERP products designed to search for data outside of their own databases. They are all being refined to provide this functionality.

7.3. Future alternatives to ERP 11 systems

The ERP II standard is not the only possible strategy for the evolution of the ERP standard. The consulting group AMR Research offers its vision of the evolution of ERP systems called ECM (Enterprise Commerce Management). The concept of ECM is based on ideas about how the latest information technology will affect the application systems and architecture of corporate systems.

According to strategists from AMR Research, one company cannot create a fully functional ERP system. To solve the problem, a radically new approach is proposed, based on the application of some kind of integration backbone, through which next-generation applications will be able to connect to the corporate enterprise system without additional integration costs. Such a backbone will be connected to external applications by a "private trading exchange" (PTX, Private Trading Exchange), through which companies will work with their counterparties.

The convenience of organizing such an implementation requires a number of serious problems to be solved by manufacturers of corporate information systems. Applications must perceive and process data created by external applications, work with the user through an interface that was created by a third-party vendor, and interact with application servers that they did not develop. In order to decide to support the ECM concept in their software products, a lot of time, resources, and costs will be required on the part of manufacturers, especially the large ones. For example, in SAP, according to AMR Research estimates, the process of code refinement and additional training of developers will take at least three years and require more than one billion dollars. However, these costs will undoubtedly pay off over time.

Despite the presence of a number of unresolved problems, the process of widespread integration of information flows is increasingly being carried out not only at large enterprises of the class of corporations, but also at a number of small companies. This shows the need to develop common standards and protocols of interaction implemented in industrial software complexes.

Several years ago, software developers and system integrators began to actively discuss problems and approaches to the creation of the next generation of ERP class solutions, which were immediately dubbed ERP III systems by the press and society [11]. IT-industry specialists and experts do not hurry to agree with this definition, and do not rush to establish any new terminology and gradation of the systems, but the need for significant changes in the structure of ERP II is expressed rather unanimously and confidently. Firstly, the number of functional components extending the standard features of ERP II should be increased. Besides SCM (Supply Chain Management System) and CRM (Customer Relationship Management System) it is possible to expect inclusion of PLM-systems (Product Lifecycle Management System), MES-systems (Manufacturing Execution System - a production process management system at the shop level), SRM-systems (Supplier Relationship Management) and IRM-systems (Investor Relationship Management). Secondly, functionality of components of intelligent analysis of business information (Business Intelligence, BI) should be extended, which will merge into one multifunctional component. OnLine Analytical Processing (OLAP) will use data from various subsystems.

It can be assumed that the development of the concept of ERP II systems will not stop here either. Previous improvements in the architecture and functionality of corporate planning and management systems give reason to expect that the generalized ERP II concept will continue to be actively expanded by incorporating new elements. The level of ERP technologies penetration into corporate environment is very high at the moment, and in order to develop business further software developers have to search for new service approaches. At the same time, best-in-class solutions, especially in SCM and supply chain management, do not yet lend themselves to the "on-demand" trend - perhaps because of the favorable (and therefore insufficiently stimulating) competitive situation.

According to many foreign analysts, BPM (Business Process Management) is now becoming one of the fastest-growing segments of the corporate software market. Today BPM claims to be the next "panacea" on which the next generation of automation systems will be based. BPM is a set of processes and applications designed to optimize business strategy implementation. That is, BPM is understood as a closed cycle of management, which includes four stages: developing strategy and determining strategic indicators of business development, creating operational plans to support the developed business strategy, monitoring the execution of operational plans and analysis of the results achieved, and regulation - adjustment of plans in accordance with actual conditions of activity and capabilities of the company.

Conclusion

Any large enterprise cannot exist without a single information system that automates all key business processes of an enterprise. Such a system is called a corporate information system, which implies comprehensive automation, i.e. translation into the plane of computer technology of all major business processes of the organization. The use of special software to provide information support of business processes as the basis of CIS seems to be the most justified and effective.

Modern business process management systems allow the integration of different software around them, forming a unified information system that exchanges the required data between the individual links. This solves the problem of coordinating the activities of individual employees and units as a whole, allowing them to provide the required information and control executive discipline, and management receives timely access to reliable data on the progress of the production process and has the means for operational decision-making and implementation of their decisions.

Most importantly, the resulting automated complex is a flexible, open structure that can be reconfigured on the fly and augmented with new modules or external software.

A modern CIS is built using a layer-by-layer principle. Thus, specialized software (office, application), document management system, document input programs, as well as auxiliary software for communication with the outside world and providing access to the system functionality via communication means (Internet, Intranet, E-Mail) can be divided into separate layers.

Among the advantages of this approach are the ability to make changes in individual software components located in one layer without the need for fundamental changes in other layers, to provide a formal specification of interfaces between layers that supports the independent development of information technology and the software tools that implement it. And the use of open standards allows the seamless transition from software modules of one manufacturer to programs of another (e.g., replacement of the mail server or DBMS). In addition, the layer-by-layer approach makes it possible to increase the reliability and fault tolerance of the system as a whole.

List of references

1. Samardak A.S. Corporate information systems: Textbook. - Vladivostok: TIDOT FEFU, 2003.   - 252 p, http://window. edu. ru/window_catalog/files/r41013/dvgu134.pdf

2. Textbook ed. by Yu.F. Telnov / Designing economic information systems / Finance and Statistics, 2003

3. O'Leary, Daniel. ERP Systems. Modern enterprise planning and resource management. Selection, implementation, operation - M.: OOO "Vershina", 2004. - 272 c.

4. Gavrilov D.A. Production management based on the MRP II standard. - SPb: Peter, 2003. - 352 p.: ill. - (Series "Theory and Practice of Management")

5. Oladov N. A., Piterkin S. V., Isaev D. V. Just in time for Russia. The practice of ERP-systems application. - M: Alpina Publishers, 2009. - 368 p.: ill. - (Series "Management Models of Leading Corporations")

6. Volchkov S.A., Balakhonova I.V. Improvement of enterprise quality with the help of ERP-class information systems (on the example of MFG/PRO), http://citforum.amursu.ru/cfin/mrp/erp_is. shtml

7. Greenberg P. CRM at the Speed of Light. Attracting and retaining customers in real time over the Internet. -M: Symbol Plus, 2006. - 528 c.

8. Payne E. The CRM Guide: The Path to Customer Management Excellence. - M: Grevtsov Publisher, 2007. - 384 c.

9. Rose, K.D. Customer-Synchronized Resource Planning (CSRP), http://citforum. amursu. ru/cfin/mrp/csrp. shtml

10. Chen E. ERP II: life after life! http://www.interface.ru/fsetasp7UrWerp/e two.htm

11. Ushakov K. Along and across the market,   http://www. cio-world. ru/analytics/369486/.

Printed on 02.02.2011

Layout - LAP LAMBERT Academic Publishing
Cover design - LAP LAMBERT Academic Publishing
proofreading - LAP LAMBERT Academic Publishing
Responsible for the issue - LAP LAMBERT Academic Publishing
Signed for publication on January 15, 2011. Size 70x100
Offset printing paper. Digital printing
Pr. 6.7
Circulation 2000 copies
Order # 32765

I want morebooks!

Buy your books fast and straightforward online - at one of world's fastest growing online book stores! Environmentally sound due to Print-on-Demand technologies.

Buy your books online at
www.morebooks.shop

Kaufen Sie Ihre Bücher schnell und unkompliziert online – auf einer der am schnellsten wachsenden Buchhandelsplattformen weltweit! Dank Print-On-Demand umwelt- und ressourcenschonend produziert.

Bücher schneller online kaufen
www.morebooks.shop

KS OmniScriptum Publishing
Brivibas gatve 197
LV-1039 Riga, Latvia
Telefax: +371 686 204 55

info@omniscriptum.com
www.omniscriptum.com

Printed by Books on Demand GmbH, Norderstedt / Germany